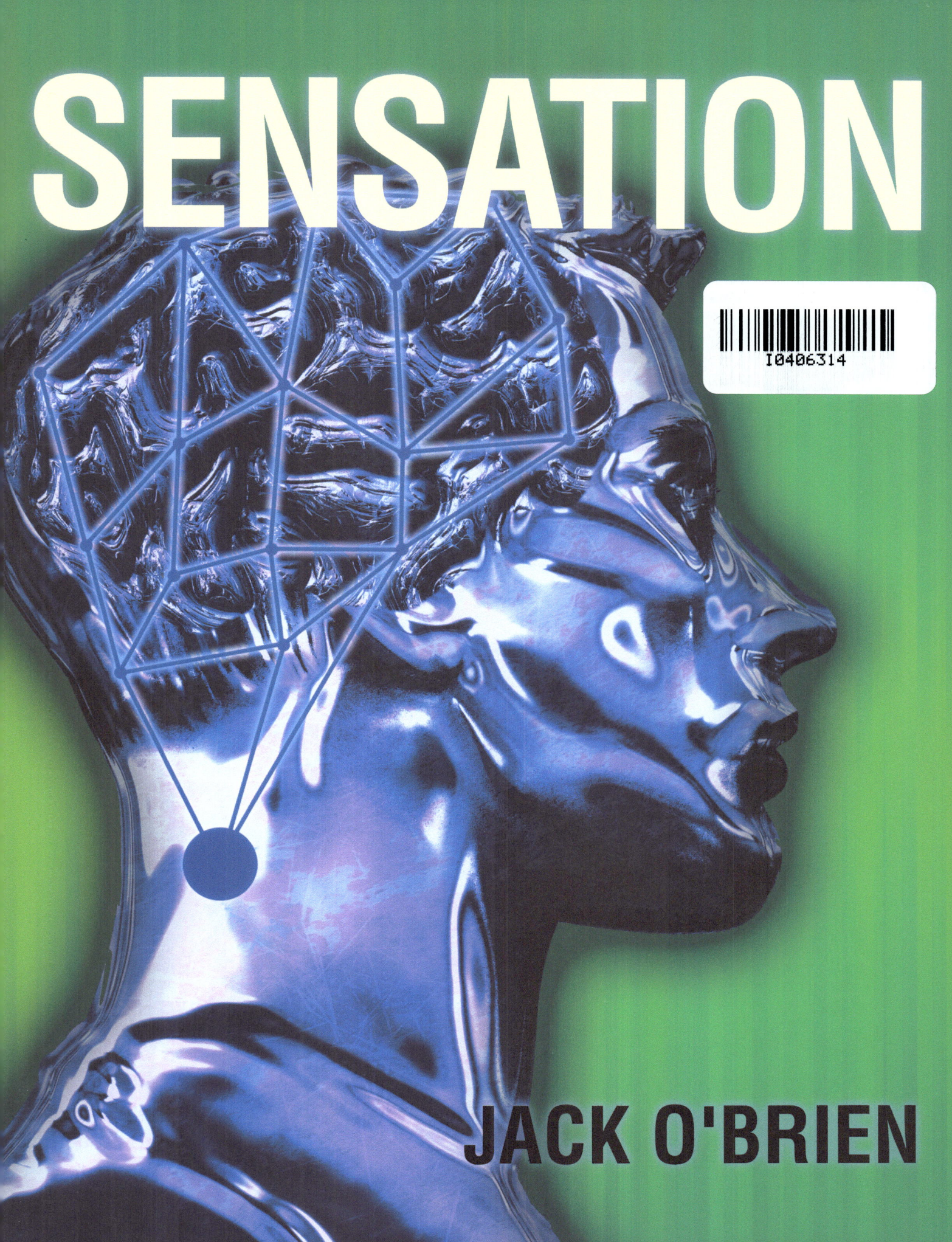

Copyright © 2014 by Jack O'Brien. 539858

ISBN: Softcover   978-1-4931-6863-7
       Hardcover 978-1-4931-6862-0
       EBook   978-1-4931-6864-4

All rights reserved. No part of this book may be reproduced or transmitted in any form or by any means, electronic or mechanical, including photocopying, recording, or by any information storage and retrieval system, without permission in writing from the copyright owner.

Rev. date: 04/01/2014

To order additional copies of this book, contact:
Xlibris LLC
1-888-795-4274
www.Xlibris.com
Orders@Xlibris.com

INDEX TO SENSATION!

	Page
SUMMARY	5
INTRODUCTION	7
CONSCIOUSNESS, SENSATION AND PERCEPTION	9
THE SENSORY NEURON	11
THE DNA MOLECULE AND PROTEIN SYNTHESIS	17
AN APPROACH TO REMEMBERING	21
MEMORY STORAGE AND RECALL	23
THE SUPRESSION OF ACTION POTENTIALS	27
SHORT-TERM MEMORY	31
EMOTION	33
QUESTIONS OF OPINION	35

SUMMARY

Bearing in mind the current problems of mass murder, veteran suicide and female depression, all of which seem to suggest some kind of breakdown in the mental health of the individuals concerned, the purpose of this monograph is to explore the workings of the human sensory nervous system, and in particular the role of memory, which plays a crucial role in directing our actions.

INTRODUCTION

All species capable of vocalizing (excepting the echo-locators) owe this faculty to the FOX P2 gene on the seventh chromosome, which regulates a series of genes engaged in setting up the vocalization muscles during gestation and connecting them to the Central Nervous System (CNS) so that the emission of sound is controllable. One of the principle changes involved in the evolutionary progression of apes into hominids was a change in two proteins in the FOX P2 gene which had the effect of substantially increasing CNS control of the vocalization muscles.

This change conferred a substantial evolutionary advantage, since it permitted a greatly increased degree of cooperation between hunters and therefore an improved shot at procreation. The change probably occurred due to a mutation around three million years ago, and it took until a mere three thousand years ago to translate meaningful speech into written characters.

But from there was no turning back. Writing allowed communication with the future instead of just with contemporaries, providing a continued accumulation of knowledge, permitting homo to become the first species able to dominate and alter rather than simply endure its environment.

Virtually all living organisms have a memory, in the sense that they avoid or withdraw from previously-encountered danger. In fact it is probably a fundamental requirement for life, since without that faculty, an organism is not likely to live long enough to reproduce itself. But access to the information recorded and stored by others represented an extraordinary advance, repeated in our own time by the free and immediate availability of all transmissible knowledge on the Web. This is a development, which at this very moment is in process of transforming society by breaking down the traditional barriers to learning.

At the conscious level, there is no mystery about learning, which of course involves memory. If an event is sufficiently intense, or otherwise repeated often enough, we shall recall it in future at will. At the nanoscale level, however, there are several difficulties. How, for example, can we recall a memory almost instantaneously; and how can we recall our major part in a play lasting two hours; how can a concert pianist or violinist recall the millions of muscle movements involved in a concerto, or an acrobat, the split-second coordination of movements demanded by the high trapeze? Anyone capable of such feats will tell you they are achieved by constant practice, but what about visual memory which we can all achieve with a minimum of effort at a moment's notice? The answer to all these questions lies in the FOX P2 gene, which simultaneously connects the formidable information storage capacity of the DNA (deoxyribonucleic acid) molecule to the CNS, and to the vocalization muscles which initiate recall.

A sufficient understanding of events at the molecular level requires a preliminary review of the various components, set forth in the following sections.

CONSCIOUSNESS, SENSATION AND PERCEPTION

The nominative process is so fundamental that we cannot discuss, or even think about something, unless we give it a name. We are endowed with a large number of sensors: the familiar five plus those responding to skeletal position, muscle stretch, orientation, inflammation and trauma, and a variety of environmental conditions. Each sensor is continually discharging into its own neuron a train of electrical signals indicating its present condition, and the sum total of all these signals we recognize as Consciousness. But Consciousness is a Sensation, and to describe what we feel, even if only to oneself, we must first give it a name, and we cannot give it a name unless we have learned the name at our Mother's knee or later, and that means the name must have been memorized. So everything we see, hear or taste, etc. is what we have learned and committed to memory as a name, and recalled from memory when we feel it as a perception.

But what is it that we recall? Obviously it is only a feeble copy of a sensation. A sensation has a feeling of immediacy; if visual, it is highly detailed, in full color and alive, but we have only to close our eyes to realize that what we now perceive is only a bad copy, without detail or movement and generally colorless. But at least we can describe it, because we have been taught to give everything a name. What we remember is called a Perception, and one purpose of this monograph is to suggest how sensations become changed into perceptions through a memorization process.

There is no great problem in demonstrating how most sensations, which are nothing but the motion of ions into and out of our nerve membranes, are stored and recovered, but vision is different. A visual scene bombards several million photocells in the retina with photons all at the same time, and there are only about a million neurons in the optic nerves able to carry all their signals simultaneously. But the visual image is constantly changing as we move our head, and although there are a hundred billions neurons in the Central Nervous System (CNS), and the output of peripheral photocells can be ignored, there is no way to store that torrent of information received over a lifetime.

To resolve the problem, Nature has arranged that what is stored is a description of the scene rather than the scene itself, and that can be stored and recalled like any other auditory/vocal signal, requiring a fraction of the channel capacity needed for the original. Which is why, when we close our eyes and describe what we have just seen, that is all we get: a description and not a photograph, and still less a movie.

Other types of memory employ the same technique. We cannot recall the identical sound of an orchestra; the best we can do is to hum the melody or play it on a piano. Similarly we recall a smell or a taste by thinking or saying "It smelled like violets" or "It tasted like pork". We become so accustomed to this arrangement that we can really convince ourselves that we actually smell violets or taste pork, but it is an illusion.

We owe this solution to the modified FOX P2 gene, which perfected the neural connections between all those sensors which generate a practically continuous signal and the vocalization musculature.

Since memory is eminently a neural process, the following section discusses the construction and operation of the sensory neuron.

THE SENSORY NEURON

The sensory neuron, meaning a neuron which is connected to a particular sensor, is an extended cell comprising the soma, an enlarged section containing the nucleus; the input dendrites which project from the soma and branch like the roots of a tree; and the axon, which also projects from the soma to a distance much greater than the input dendrites. The axon may also terminate in a small number of dendrites, called the exit dendrites.

The outer membrane of the neuron is bridged by a large number of channels which (separately) permit the entry and exit of sodium and potassium positive ions when required. This membrane encloses an electrically-conductive solution called the cytoplasm. The nucleus is enclosed by its own membrane, which is also traversed by a number of channels which permit the passage of whole molecules in addition to ions, and which contains the nucleoplasm.

Like virtually all cells, the nucleus harbors the 24 duplicated chromosomes which carry the Genetic Code, each housed in a separate DNA (deoxyribonucleic acid) molecule wrapped around a histone structure for the purpose of compaction. As distinct from other cells, however, the neuron employs the genetic code only to synthesize proteins for its own use but not, with minor exceptions, to replicate itself.

The principal purpose of the sensory neuron is to transmit signals from a sensor, such as a photo-sensitive retinal cell or a resonant hair in the cochlea of the ear, to a muscle, a gland or to other neurons which participate in sensation, action, memory, decision and so on. Each sensor responds to its particular stimulus by delivering a signal, either chemical or electrical, which initiates an Action Potential (AP) in the neuron.

Signals in the form of action potentials are passed on from one neuron to the next across a Synapse. Every dendrite is tipped with a small membrane which on the transmitting side is called a Pre-Synaptic Membrane, and on the reception side is called a Post-Synaptic Membrane. The two membranes butt up against but do not touch one another, the narrow space between them being called the Synaptic Cleft. Transmission of signals across the Cleft is undertaken by chemical neurotransmitters, molecules of which are synthesized in the Pre-Synaptic side and dock on the Post-Synaptic side. The purpose of this arrangement is to permit chemicals in the bloodstream to penetrate a very large number of clefts at the same time and either inhibit signal transmission, as in sleep or coma, or accelerate transmission when urgent action is called for. The region of the neuron immediately downstream of the soma, called the Hillock, is particularly rich in synapses connecting the neuron to the dendrites of neighboring neurons.

A small proportion of synapses are known as electrical rather than chemical synapses, permitting direct passage of APs from one neuron to the next, presumably for ultra-rapid emergency response or to evade the Threshold Potential (see below) when necessary.

The cytoplasm of the neuron is normally maintained at an electrical potential, called the Resting Membrane Potential, of about 70 millivolts (mV) negative with respect to the exterior of the neuron. This automatic procedure involves the independent movement of sodium and potassium ions into or out of the cytoplasm to maintain the resting membrane potential at -70 mV, as necessary. The resulting adjustments to the resting membrane potential are relatively small, only a few mV, compared with the very much larger Action Potentials involved in signal transmission.

The response of most sensors to stimulation is to deliver into the cytoplasm of its neuron, either directly or via a chemical neurotransmitter, a positive electrical charge sufficient to depolarize the neuron, meaning that it drives the local membrane potential in a positive direction to around 40 millivolts negative to the exterior. Immediately, the voltage-gated sodium channels open wide to admit a rush of positive sodium ions, driving the membrane potential as much as 30 mV into the positive region, whereupon the local potassium ion channels open and expel positive potassium ions into the exterior, bringing the membrane back to slightly less than the resting potential, which acts to prevent backward transmission of the signal pulse. The positive potential pulse lasts about one millisecond, and the recovery period during which a new pulse cannot be transmitted is about twice that. The potential 'spike' is termed an Action Potential, and the neuron is said to have 'fired'.

It may be questioned that the foregoing description apparently leaves a surplus of sodium ions and a deficit of potassium ions in the cytoplasm which, after several firings could disable the neuron. The literature is silent on the matter, other than to state that "surplus sodium ions are rapidly expelled from the cytoplasm", so that the preceding description remains tentative. Nevertheless, the existence of Action Potentials is experimentally factual.

At any rate, as the potential falls towards the resting membrane potential, the next sodium channel in line opens wide and admits a flood of positive sodium ions further along and repeats the process, which, once started, is self-sustaining.

The signals emitted by a sensor normally consist of a rapid train of spikes separated by some two milliseconds defining the intensity of the initial stimulus, followed by a series of more widely-separated spikes, which by their repetition rate identify the sensor from which they came. Depending on conditions at the synapses, signals can be propagated at a speed of up to 120 meters per second.

Each signal pulse continues on towards the nucleus and the axon, though it is important to note that there is no charge or particle movement along the whole length of the neuron, but only a handing-on of energy from one ion channel to the next, thus avoiding the attenuation inseparable from continuous longitudinal movement.

As the signal progresses towards the axon, it will stimulate numerous synapses along the way, initiating Action Potentials in their corresponding neurons.

It is particularly important to synchronize signals from different types of sensor, particularly visual, vocalization and auditory. In this respect it seems likely that the denditric nature of the neural input structure is specifically designed for that purpose, connecting several types of sensory outputs together so that a given sensation is actually a combination of several different sensory signals. Structures such as the Hippocampus, the Amygdala, the Thalamus and the Nucleus Accumbens, which bundle many neurons closely together to facilitate 'crosstalk' between them, provide a similar function. But just as the ear is able to distinguish the sound of a single instrument within the total sound waveform of a complete orchestra, so composite action potentials can be recognized separately following analysis by the corresponding neural networks.

A very necessary culling of non-significant stimuli, which constitute perhaps 99% of the total, is performed by the Threshold Potential. This is the minimum change in the resting potential, generally around 30 mV towards the positive, which will cause the neuron to fire. Insufficiently intense stimuli are simply ignored, although there are various stratagems designed to draw attention to potentially dangerous situations, such as the unconscious saccadic movement of the eyeball, which momentarily projects peripheral images onto the macula. In general such stratagems are reflexive, meaning that muscular action is achieved instantly without submission to CNS decision and control.

The threshold potential is not constant, but can be reduced following unusually frequent APs, or even increased following lengthy disuse. In a process called Synaptic Plasticity, reduction in the threshold potential in turn gives rise to Long-term Potentiation, which defines a preferred pathway for a given stimulus and accounts for the network properties of decision-making, reasoning, the fight-or-flight reflex, and so on.

An enormous proliferation of neurons occurs in very early childhood, when the infant is rapidly learning to convert sensations into perceptions: to walk, talk, and reason. Many of these excess neurons will die before puberty, suggesting that their function was to coordinate reactions that subsequently became reflexive. Following puberty, however, most neurons do not replicate. It is tempting to speculate that outstanding intelligence, eidetic recall, etc. may be due to unintended retention of excess neurons, though the advantage they confer is often outweighed by behavioral problems.

The following section discusses the DNA molecule which, in the sensory neuron only, is primarily engaged in protein synthesis, and possibly provides memory storage.

THE DNA MOLECULE AND PROTEIN SYNTHESIS

The DNA molecule consists of two separate strands, here called the A and B strand, held together by hydrogen bonds. Each strand is a helix, comprising a succession of sugar and phosphate molecules forming a 'backbone', with each sugar molecule attached to a nucleotide such as adenine (A), cytosine (C), guanine (G) or thymine (T). The order in which nucleotides are attached along the backbone constitutes the Genetic Code, thus one might encounter AGTCAAGTATCCTTGA etc. as one travels from head to tail of a strand. The two helical strands intertwine in opposite directions, with adenine on the A strand always pairing with thymine on the B strand via two hydrogen bonds, and cytosine on the A strand with guanine on the B strand via three hydrogen bonds. The individual nucleotides are known as Bases, and when joined are called Base Pairs. The purpose of this arrangement is twofold: firstly to provide the binding units or Codons taken three at a time, such as AGT, from the coding sequence and inserted into a variety of amino acids to constitute proteins which make up both the body and its regulatory functions; and secondly, in reproduction, to convey into every cell in the new organism the whole Genetic Code, made up of strings of associated codons known as genes, which replicate the same structure and function as in the parent.

There are only some 32,000 genes in the human Genetic Code or Genome, but between them they occupy over 3 billion base pairs, which is too much for even the DNA molecule to carry, so the genes are split into 24 separate structures called chromosomes, each containing anywhere between 46 million and 247 million base pairs of nucleotides. Each chromosome is allocated to a separate DNA molecule, so that there are 24 DNA molecules in every nucleus. The Genome itself is made up of protein-coding nucleotides sequences, known as Exons, widely separated by lomg stretches of non-coding sequences, known as Introns. Each of the two strands of DNA carries a duplicate of its particular chromosome, though the order of nucleotides in one strand is the reverse of the order in the other.

There may be as many as 250 million base pairs in a single DNA molecule, which is always confined to the nucleus of every cell in the body. To fit the long DNA molecule into the nucleus, it has to be wound around a framework of eight linear 'sticks' of the protein histone. Histones are not bound to DNA but act as a mould to prevent the DNA from unraveling and to deny access, by expanding radially, to the many enzymes in the nucleus until they are needed. In non-neural cells, DNA fulfills the double function of initiating the synthesis of proteins for cell maintenance, repair and communication with other cells, and cell replication, which can be as often as every few days in gut and skin cells. Neurons, however, do not normally replicate after puberty, and so for them protein synthesis is the primary duty of DNA, although it seems possible that they also provide memory storage.

Normal operation of any cell involves a continual loss of highly-structured protein molecules when they have completed a specific task and are allowed to disintegrate into their component molecules. To replace these, DNA undertakes a process called transcription followed by translation. On receiving a chemical signal to replace a particular protein, the histone framework relaxes to make the coils of DNA over a stretch of about a dozen base pairs available to the transcription enzyme RNA polymerase. Relaxation starts at a specific codon sequence, known as a Promoter, upstream of the required stretch of DNA, and stops at another sequence called a Terminator.

At the same time the enzyme helicase breaks down or 'unzips' the hydrogen bonds between the base pairs over the selected stretch of DNA, and the enzyme RNA (for ribonucleic acid) polymerase moves in and, using the exposed B strand as a template, synthesizes a new single-strand molecule known as pre-mRNA, or pre-messenger RNA, in which the thymine nucleotide is demethylated to form uracyl. This change is necessary because DNA is a hydrophobic molecule which cannot pass through the membrane pores of the nucleus, while RNA is a hydrophilic form exportable into the cytoplasm.

RNA polymerase, which controls the helicase enzyme, unzips further stretches of DNA in order to manufacture several additional short single-strand RNA molecules known as tRNA, for transfer RNA, and rRNA, for ribosomal RNA, together with several others used for transcription and translation purposes.

Immediately following RNA polymerase synthesis of each succeeding base on the B strand, the enzyme DNA polymerase moves in and repairs the broken hydrogen bond, zipping up the strands and re-tightening the histone structure.

Still in the nucleus, the enzyme spliceosome cuts out only the required exons, or protein-coding nucleotides, which may be located at widely-separated sites, from the pre-mRNA molecule, joins them together, and ships the product, now known as mRNA, out of the nucleus into the cytoplasm. The abandoned introns, or non-coding nucleotides, are presumably picked up by RNA polymerase for subsequent synthesis operations. The tRNA, rRNA and miscellaneous regulatory molecules are shipped out at the same time.

Once in the cytoplasm, the tRNA molecule binds to two amino acid molecules for subsequent incorporation in a polypeptide, and a ribosome, highly complicated assembly of RNA and several proteins, having the ability to move along the successive bases of a nucleic acid polymer, couples to both the mRna and tRNA and ejects a polypeptide string of connected amino acid and mRNA codons which are subsequently spliced into separate proteins. There are only some 20 proteins having distinct chemical compositions, but there are thousands of different protein structures which owe their activity to the way in which they are three-dimensionally folded for docking with their target molecules. The particular folding required is performed both during and after the cutting process.

Complicated as it may seem, the foregoing is only a bare summary of the basic steps in protein synthesis, omitting many details of the processes by which the various enzymes and helper molecules bind and release. Even so, it gives some idea of the crowded nature of both the nucleus and the cytoplasm and the manifold possibilities for error.

At this stage it seems advisable to point out that the DNA molecule, remarkable as it is, has limitations. In particular, helicase cannot unzip the full length of the molecule at one time, and must pause after maybe a dozen base pairs before re-initiating the histone relaxation and unzipping processes. This has some importance when considering the possible role of DNA in storing action potentials, and may be responsible for our sense of rhythm and cadence in music, song, prose and poetry.

AN APPROACH TO REMEMBERING

Psychologists recognize at least five distinct types of memory: Episodic, for unique events such as an accident; Declarative, for verbalized recall such as a recital; Procedural, for practiced actions such as playing tennis; Working, for numbers and letters in daily use; and Spatial, for finding our way around.

Unquestionably, any convincing theory of memory must account for every one of these manifestations. But we should not be misled into thinking that each of these implies a different mechanism of storage and recall, any more than we would conclude that the various symptoms displayed by a single disease each imply a different disease. Surely there must be a single basic mechanism which, depending on the circumstances in which it is applied, will explain all the different manifestations observed.

Thus this monograph takes the view that all those of our different sensations worthy of long-term recall are integrated in the CNS in the same way as the individual sound of a single musical instrument is integrated in the overall waveform of sound generated by a whole orchestra, yet is separately distinguishable by the ear. Not surprisingly, then, the integrating factor is sonic, because we are exquisitely capable of analyzing sonic frequencies.

The first step in this synthesis is the proposition that all sensations are registered and recognized as ionic neural impulses called Action Potentials (APs) and that these are interpreted vocally or sub-vocally in descriptive terms through a learning process accomplished in early childhood. The next step is to propose that APs are physically aggregates of sodium ions which can be permanently stored on any one of the DNA molecules located in the nucleus of every sensory neuron.

Next, that in the normal process of protein synthesis, the stored ions are released into the cytoplasm of the neuron so as to establish APs in their own right. And lastly, that such release takes place either haphazardly during normal degradation of the carrier molecule within several hours or deliberately by a cueing process which releases all of the ions at one time and in their original order.

The cueing process is perhaps the hardest to understand, as it assumes that a sub-vocalized or vocalized cueing signal, having some association with a stored sensation will somehow recognize its associate in the cytoplasm. And that recognition will trigger immediate detachment of stored ions from their carrier molecule and in their proper order. There are, in fact, dozens of chemical processes in every cell which constantly perform this type of operation, so that the basic process is not in question, but its specific application in the neural cytoplasm is not clear, and a detailed explanation will have to await experimental confirmation.

The following section goes into more detail concerning the proposed memory storage and recall process.

MEMORY STORAGE AND RECALL

A sensation only qualifies for memory storage if it has sufficient intensity to enter the CNS, meaning that it exceeds the threshold potential and gives rise to an Action Potential (AP) sufficiently strong to propel a few ions through the nuclear membrane. In detail:

The passage into the nucleus of the first sodium ion of an AP signal train provokes the same reaction as a signal calling for protein synthesis, except that following histone relaxation helicase starts unzipping the DNA double helix at the first promoter code sequence rather than at the specified protein coding interval, which might be several thousand base pairs away.

The sodium ion then bonds electrostatically to the nearest nitrogen atom of the phosphate backbone group of the first intron, or non-coding nucleotide sequence, that it finds. The enzyme DNA polymerase promptly restores the broken hydrogen bonds to seal the ion in place. RNA polymerase does not attempt to copy the exposed B strand, because it has not received a protein-coding signal.

If a new sodium ion arrives a couple of milliseconds later, it docks on the next phosphate segment of the DNA backbone. If no ion arrives, signifying a zero in computer digital code language, DNA polymerase nevertheless restores the hydrogen bonds of that segment to seal the exposed backbone and preserve on the DNA molecule the space and time sequence of the originating APs in the cytoplasm. Concerning the synchronization of APs and the entry of sodium ions into the nucleus, it should be observed that the ion channels in the nuclear membrane admit sodium ions at the rate of one every millisecond, corresponding to the maximum frequency of APs in the cytoplasm. The sodium ion bonds to, and thereby ionizes, only the phosphate group of any DNA B strand segment, and only on non-coding introns, since coding segments, or exons, carry a positive charge which repels the positive ion.

The storage process continues until either the AP signals cease or the whole DNA molecule is fully zipped-up by DNA polymerase. In the extremely rare case where APs continue to arrive after the whole storage capacity of a single DNA molecule is reached, storage continues on the next DNA molecule, although the delay involved may cause problems of recall. It is possibly for this reason that theatrical performances have several acts and musical compositions several movements, allowing performers to use only one DNA molecule at a time.

On termination of the AP sequence, the histone structure reverts to its constricting condition and the DNA molecule is said to be 'loaded'; the signal composed of ionized non-coding segments on the DNA B strand will remain there indefinitely. The next time the nucleus receives a call for any one of 20 different proteins, the protein synthesis procedure previously described commences and continues through the phase of re-joining the previously separate exons by spliceosome. This time, however, the enzyme inectase (another variant of the ribosome family), recognizing the presence of ionized introns, reassembles all the introns discarded by spliceosome, cuts out only ionized introns and joins them together to form the single-stranded molecule iRNA, discarding the non-ionized introns for degradation in the cytoplasm.

There is a limit, probably around 500 bases, to the length of an iRNA molecule that can be exported into the cytoplasm at one time and therefore inectase, on reaching the limit, expels that quantity into the cytoplasm, shifts to another molecular transport pore and expels an equal or lesser quantity of iRNA, depending on the number of ionized segments involved, into the cytoplasm.

Within the cytoplasm a ribosome, failing to recognize a protein-coding exon, cannot latch on to the iRNA molecule for peptide synthesis. But there is a variant of spliceosome in the cytoplasm which ligates or joins together successive lengths of iRNA as they issue from the nucleus, to form a complete iRNA single-strand molecule carrying the whole message. This molecule remains in the cytoplasm either to degrade into its component parts, releasing its sodium ions in a haphazard and incoherent manner, or, on recognizing a cueing signal, discards all of its sodium ions simultaneously in the same temporal and spatial sequence as they existed in the originating APs. In the former event they are simply included in the inrush of sodium ions during the next series of APs, but if coherent and instantaneous following cueing, they are recognized as legitimate memories. They are generally too attenuated to constitute Perceptions, but on reinforcement by identical APs from neighboring neurons, they become full-blown memories.

Thereafter, the whole process will be repeated every time protein synthesis is called for, irrespective of the protein specified. As previously mentioned, protein synthesis is a virtually continual process, so that a cueing signal can always be assured of a target. Indeed, a loaded iRNA molecule in the cytoplasm may easily be cued by a stray thought or the sound of a word, and instantly drops its ionic charges to form a memory. It is not difficult to attribute dream material to this phenomenon, though it is more likely that the irrational and often inconsequent nature of dreams occurs due to short meaningless APs arising from the disintegration of uncued RNA in the cytoplasm. To improve that probability of memory recall, the original signal may be repeated several times so as to increase the number of neurons involved.

A cueing signal is a short sequence of APs identical or very similar to a sequence in the original signal. Where the cueing sequence in its progress through successive neurons chances to correspond with the same sequence of charges bonded to an intron lying in the cytoplasm, the sum of all charges within the cueing period is significantly more than the sum of charges induced by any different, non-cueing, sequence. This increased total charge intensity triggers a ribosome-type polymer crawler enzyme to latch on to the intron and proceed along its length, detaching successive positive charges into the cytoplasm in the same order as originally established on the DNA molecule in the nucleus.

There is some question whether the released charges are sufficiently intense to establish a perception, and perhaps the purpose of the electrical synapses is to permit a buildup of charge density as the sequence joins that of neighboring neurons.

THE SUPRESSION OF ACTION POTENTIALS

The evanescence of organic material requires that organic cells have at least two basic duties: to execute the particular function for which they were created, including the ability to synthesize proteins for structural maintenance and repair, and secondly, to replicate when necessary to build new tissue or replace the whole cell when wear and tear pass the point of repair. Unfortunately the complex electrochemical processes involved in protein synthesis are incompatible with the delicately timed process of cell replication, and the solution reached by Evolution has been to separate the two by the intervention of sleep. In sleep, the continuing flow of positive electrical charges into and out of the cell in order to maintain the interior potential at around 70 millivolts negative relative to the exterior is halted, allowing protein synthesis to proceed undisturbed, while on waking the process of replication can proceed without the interference of enzymes essential for protein synthesis.

This division of labor is particularly important to neural cells whose basic function is to transmit messages involving relatively huge and rapid ion transfers back and forth across the cell membrane to establish Action Potentials as high as 120 millivolts across the membrane, which is totally incompatible with protein synthesis. Actually, cell replication is incompatible with both messaging and protein synthesis, so sensory neurons have largely relinquished the power to reproduce after their enormous increase in early childhood to assist the learning process, and so have developed proliferating dendrites interconnected with neighboring neurons by a host of synapses so that the death of a single cell scarcely affects their combined function.

In point of fact, a sleeping period for all types of cell is effected by the injection of hormones into the bloodstream which penetrate the Synaptic Cleft of all sensory neurons in about a quarter of the brain at one time. These hormones either block the synthesis of neurotransmitter in the Pre-Synaptic membrane of each Cleft, or block the receptivity of receptors in the Post-Synaptic membrane, depending on location. Normal messaging promptly ceases, and that quarter of the CNS becomes unconscious. Of course, reflex activities such as heartbeat and breathing, which operate independently of CNS control, remain unaffected. The cessation of messaging in only a fraction of the CNS at one time, detectable by EEG readouts, is apparently a security measure so as to assure that three quarters of the CNS is available for fight or flight at any time. Normal awakening occurs when the production of sleep hormone is discontinued and synaptic transmission resumes automatically. Emergency awakening requires the synthesis of a cleanup hormone which reverses the action of sleep hormone.

An entirely different though equally important function of the neural synapses is the suppression of unwanted or superfluous Action Potentials. Social animals have to guard against anti-social behavior, so every human society, irrespective of creed or custom, has developed a code, similar to the Mosaic Code, listing 'prohibited' behavior, particularly including murder. This Code is instilled in the memory of every child from the earliest years and provides the yardstick against which every proposed action is measured. Aside from social issues are the questions of self preservation, involving the assessment of risk and feasibility, learned at the same time.

An action is a long sequence of muscular contractions and relaxations resulting from APs sequences practiced and committed to memory while in utero and early childhood. When initiated later, it progresses without continuous CNS control until consciously terminated. It can be disconcerting to learn that an action is initially proposed in the CNS below the level of consciousness several milliseconds before muscular activity commences. What happens in that short interval is that a CNS decision center constructs several alternative action scenarios and submits each for comparison with the stored memory sequences lying in the cytoplasm. Of course, this operation involves creating APs, which must be promptly suppressed by synaptic blocking before they initiate inappropriate or even conflicting muscular actions. Precisely to avoid potential conflict only one 'approved' scenario is permitted to proceed unsuppressed towards the muscles, and rises to the conscious level only when physical action commences.

Of all mechanisms conserved by Natural Selection, this censorship process is undoubtedly one of the most questionable. For a variety of reasons, including poor moral education, unavailability of the appropriate 'prohibit' memory in the cytoplasm at the right time, insufficient neurotransmitter because of overuse or disease, and so on, undesirable scenarios may avoid suppression and proceed to initiate undesirable muscle action typical of bad behavior. If morally reprehensible, the cure is usually re-education, but there is no cure for what is usually dubbed 'poor judgement'. Another example of tragic consequences is that in wartime, members of the armed services are deliberately trained to eliminate the designated enemy, in conflict with the prohibition against killing instilled in childhood and constantly reenforced in criminal court reporting. On return to civilian life the conflict between the two opposite instructions leads at least to deep depression and at worst to suicide

SHORT-TERM MEMORY

It is important to recall that long-term memories are stored during the waking or messaging period but, being a part of the protein synthesis process, cannot reappear in the cytoplasm as ionized introns until the following sleep or protein synthesis period. In other words, a given memory cannot be both stored and recovered in the same waking period. Yet obviously the need exists to recall activities taking place only minutes ago.

Following the production of polypeptide protein precursors by ribosomes, non-ionized mRNA molecules are released to disintegrate in the cytoplasm. The positive sodium ions which constitute APs are perfectly capable of ionizing the negatively-charged phosphate backbone of these abandoned molecules, and of responding to subsequent cueing signals which release their coherent sequence of charges to form a meaningful Perception.

The drawback to this mechanism is that the abandoned mRNA molecules degenerate into their component residues by enzymatic attack within a few hours, leaving their charges to be swept up in the wholesale movement of sodium ions out of the neuron during the messaging process. Thus while long-term memory sequences are safely housed in DNA molecules for a lifetime, short-term memories rarely survive more than a few hours. Sleep is sometimes said to consolidate memory, but while it certainly employs the protein synthesis mechanism to seal memory-candidate APs into their DNA carrier molecule, it does nothing to consolidate short-term memory sequences.

EMOTION

Emotions are hard tp pin down. There is not even general agreement on their names; those considered here are: Anger, Disgust, Fear, Happiness, Sadness and Surprise, but there are many different proposals. Generally speaking, an Emotion is a Sensation induced by strictly chemical means, usually a hormone, as distinct from a neural Sensation. There are more than fifty different human hormones, all synthesized in specialized glands, some being distributed interstitially and therefore locally while others reach every part of the body in the bloodstream. The only justification for including Emotions in a study of Sensation is that they are a type of Sensation and if distributed in the bloodstream can interact with the CNS via the Synapses and influence behavior by modifying AP transmission over a wide range.

It may be assumed that Emotions are under CNS control, though this is not necessarily or invariably so. And the question is how. One possible explanation is that given the interconnection between neighboring neurons via Synapses , the APs passing through a given neuron are likely to originate in more than one distinct Sensor. This does not corrupt any of the mixed messages, since the target gland, muscle or cortex is tuned to respond only to its assigned Sensor, but it does mean that certain structures in the CNS, such as the Amygdala, can respond to a mixture of messages and emit its own characteristic message to a hormone synthesizing gland.

On this reading, an Emotion may be regarded as the body's reaction to two or more simultaneous neural Sensations. For example, unexpectedly coming upon a sleeping tiger, involving a range of undesirable sensations, is likely to first provole Fear and induce massive secretion of adrenaline to facilitate rapid withdrawal. Conversely, the prospect of a good meal, provoked by sight, smell and possibly taste, is likely to induce a feeling of Happiness, while the loss of a companion induces a feeling of Sadness, and so on. All this is entirely fanciful, since aside from adrenaline production the Evolutionary advantage of Emotion is not always obvious.

One interesting observation, however, is that the interaction of the two types of Sensation may create a positive feedback. Hormone stimulation initiated in the CNS affects AP activation in the Synapses, which presumably in turn increases hormone production, making an emotional reaction much more intense than a run-of-the-mill reaction, though here again the Evolutionary advantage is no always obvious. Perhaps it is best not to stray too far into this area, which clearly suffers from a shortage of established fact.

QUESTIONS OF OPINION

At the present time, the prevailing academic opinion concerning where memory is stored is that memory is an intrinsic property of the vast neural network, comprising some 100 billion neurons, each having an average of about one thousand synapses connecting it to its neighbors. This opinion is largely based on the work of Donald O. Hebb, published in 1949, and the enormous success of Frank Rosenblatt's Perceptron, now found at every supermarket's checkout counter. The Perceptron was a pattern-recognition device modeled on the human brain, consisting of an electrical network with internal connections of variable strength corresponding to neural Synapses. Added support for the network hypothesis is provided by the universal conviction that anything is possible if the numbers are large enough.

The search for an alternative was provoked by the realization that a mere 100 billion neurons were totally inadequate to serve the sensory needs of maybe 80 trillion body cells together with the CNS requirements for decision-making and the storage of memories accumulated over a lifetime. An obvious solution was provided by the 24 chromosomes carried on each of the two strands of the 24 DNA molecules permanently lodged in the nucleus of every one of the Central Nervous System neurons. These 24 carry between them the whole genome of 3 billion base pairs of nucleotides in less than 2% of their available capacity, leaving the function of the remaining 98% unexplained. The arithmetic is convincing, though the precise storage and recovery mechanism is, pending much laboratory research, simply a matter of opinion.

Fortunately or not, the true location of memory is not presently a matter of commercial concern, though it may be within a few more decades, and funding for experimentation is correspondingly inadequate. In this situation, the best we can offer is opinion to guide further laboratory work.

www.ingramcontent.com/pod-product-compliance
Lightning Source LLC
Chambersburg PA
CBHW050405180526
45159CB00005B/2158